Water
75

竹子的馈赠

Generous Grass

Gunter Pauli

[比] 冈特·鲍利 著

[哥伦] 凯瑟琳娜·巴赫 绘

隋淑光 译

上海远东出版社

丛书编委会

主　　任：田成川

副主任：何家振　闫世东　林　玉

委　　员：李原原　翟致信　靳增江　史国鹏　梁雅丽
　　　　　任泽林　陈　卫　薛　梅　王　岢　郑循如
　　　　　彭　勇　王梦雨

特别感谢以下热心人士对童书工作的支持：

匡志强　宋小华　解　东　厉　云　李　婧　庞英元
李　阳　刘　丹　冯家宝　熊彩虹　罗淑怡　旷　婉
杨　荣　刘学振　何圣霖　廖清州　谭燕宁　王　征
李　杰　韦小宏　欧　亮　陈强林　陈　果　寿颖慧
罗　佳　傅　俊　白永喆　戴　虹

目录

Contents

在树木稀疏的小

山坡上，一只貘正看着一只松

鼠在树木间跳来跳去。

"你还记得这些山上曾经长满了竹子

吗？"貘问道。

"我没看到过，但是我祖母曾经说过。她

还说那时候这里更凉爽，每年的降雨量

更大。"松鼠回答道。

On
a sparsely wooded
hillside, a tapir is watching a
squirrel leap from tree to tree.
"Remember when these mountains
were covered in bamboo?" asks the
tapir.
"I never saw it myself, but my grandma
told me about it. She also said that it
used to be much cooler during the
day and that we had more rain
all year round," replies the
squirrel.

山上曾经长满了竹子

Mountains were covered in bamboo

桉树种得越来越多

Eucalyptus tree became so popular

"嗯，自从桉树

种得越来越多，土地就变得越

来越干燥了。"

"为什么人们要种这种疯狂吸水的树，

并且放任它生长呢？这对土地一点好处也

没有。"

"你知道，农民们听说这种外来的

树长得比其他树更快。"

"Well, since the
eucalyptus tree became so
popular, the soil is much drier."

"How can anyone plant a tree that
guzzles so much water? And then
allow it to keep on growing? It's not
good for the soil."

"You see, farmers were told that
this alien tree would grow faster
than any other tree."

"甚至比竹子长
得还快？种子公司肯定有一位
出色的推销员！"
"是的。你知道种上竹子后，多长时间
才能收获一根25米长的竹材？"
"我听说只要3年就可以。"松鼠回
答道。
"那么再次收获需要多长时
间呢？"

"Even
faster than bamboo?
The seed company must
have had a excellent salesman!"
"Right. And do you know how
long after planting it you will be able
to harvest a twenty-five-metre tall
bamboo?"
"I hear you can do so in just three
years," the squirrel replies.
"And when can you harvest
it again?"

3年后就可以收获竹材

Harvest bamboo in just three years

9

Heliconias

"第二年就可以。"

"没错，更重要的是，竹子可以连续收获70年以上。在这个时间段内，它们产生的纤维素是生长最快的树种的60倍。另外，它们还是最美丽的花——海里康花的寄主！"

"是啊，海里看花真是美妙无比。"

"The following year."

"Exactly. And what's more, is that bamboo plants can be harvested for up to seventy years. During that time, they produce sixty times more cellulose than the fastest growing tree in the world. Plus, they play host to the most beautiful flowers: heliconias!"

"Ah yes, those helicopters are gorgeous."

"不是海里看花，
是海里康花！听着松鼠，我们
要更深入地了解竹子。"

"你说得对。树木被砍伐以后还要重新种
植，而且它们不断地消耗水，而竹子却能
保护水土不流失，除了关爱以外，竹子不
需要其他任何东西。"

"你知道谁住在高地上吗？"
松鼠问道。

"Not
helicopters, heliconias!
But listen here, Squirrel, we
need to find out more about
bamboo."

"You're right. Trees need to be
replanted after harvesting and they
need water all the time, while bamboo
can hold water and needs nothing
else, except some loving care."

"Do you know who lives in
the highlands?" asks
the squirrel.

除了关爱以外，竹子不需要其他任何东西

Needs nothing else, except some loving care

种豆子的农民

The bean farmers

"种豆子的农民。"

"对，所以这种25米长的草顶端的3米对他们很有用，可以搭豆秧支架。"

"草？"貘吃惊地问道。

"没错，竹子是一种草，所以它被收割后还可以再长出来，就像你的草坪一样。"

"The bean farmers."

"Exactly, so the first three metres of the twenty-five-metre-long grass can go to the bean farmers, who can use them as posts."

"Grass?" asks the tapir, surprised.

"Yes, grass. Bamboo is a grass. You cut it and it grows again, just like your lawn."

"再切下3米可以做竹竿，种豆子的农民用他们的毛驴把竹竿运下山，用来盖房子。"

"盖那种能随着地壳运动的节奏起舞的神奇竹屋？我喜欢这种房子。"

"剩下的东西可能有19米长，它们是很好的造纸材料。"

"Poles can then be cut from the next three metres of bamboo and the bean farmers can bring them down the mountain on their donkeys. These can be used to build houses."

"Those incredible houses that can move – dancing to the rhythms of the Earth? I love those."

"And whatever is left over, and that can be as much as nineteen metres, is good for making paper."

会动的房子

Moving houses

竹林里更凉爽

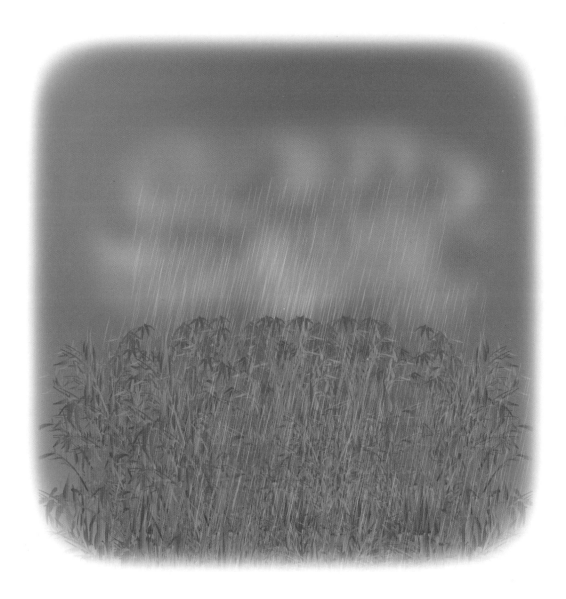

Bamboo forests are much cooler

"造纸？太神奇
了！竹子可以用来建造既舒适又
成本低廉的房子，还能创造就业机会，
让农民安居乐业。"
"它甚至能让这座山感到快乐。"貘笑着说。
"为什么？"松鼠很好奇。
"因为竹子能在下雨后储存雨水，还因为
竹林里更凉爽，云朵乐意在这里洒
下雨滴。"

"Paper
to write on? That is
amazing! Bamboo can be
used to build good, inexpensive
houses, and it creates work and keeps
the farmers happy."
"Even the mountains will be happy,"
says the tapir, smiling.
"Why?" the squirrel wants to know.
"Because bamboo holds water after
it has rained. And because bamboo
forests are much cooler, clouds
are happy to let their
raindrops fall there."

"这样我们就有更

多的雨水了！土地也能储存更多

的水！你们貘也会因此更快乐吧。"

"竹子不仅能储存水，还能滋养土地。

土地是最重要的生命之源。有了富含水分

的土壤，就意味着有足够的营养丰富的

食物！"

……这仅仅是开始！……

"So, we get more rain and the soil retains more water! Now that should make you tapirs happy as well."

"Not only does bamboo hold water, it also nourishes the soil. And soil is the most important source of life. With good soil and plenty of water, there will be an abundance of nutritious food!"

... AND IT HAS ONLY JUST BEGUN!...

······这仅仅是开始！······

...AND IT HAS ONLY JUST BEGUN! ...

Bamboo buildings are quite earthquake resistant, because they move (dance) along with the rhythms of the earth.

竹建筑物有很好的防震性能，因为它们能随着地壳运动的节奏晃动（跳舞）。

Bamboo is a grass; when cut, it grows again. Unfortunately, in Colombia, bamboo has been classified as a tree, so a special permit is required in order to cut it down.

竹子是一种草，砍掉后能再生。不幸的是，它们在哥伦比亚被视作树，只有获得特殊许可后才能砍伐。

$\mathcal{B}$amboo
grows to maturity
in only three years, and
after being cut, grows to the
same height again within a year.
It keeps on growing for 70
years. It is the most efficient
producer of cellulose in
nature.

竹子只要 3 年就能成材，砍伐后在 1 年内又能长到同样高度。它能持续生长 70 年，是自然界中最高效的纤维素生产者。

$\mathcal{W}$hen bamboo grows
in small, localised clumps, it
holds water and regenerates
topsoil. It grows in symbiosis
with heliconias and
arbolocos (also called
"crazy trees").

竹子是一簇簇丛生的，它能储存水分，促进表层土壤再生。它与海里康花及蒙坛菊（也叫"疯狂树"）共生。

When cultivated, eucalyptus grows with a fast harvesting and planting rotation cycle. Consequently, the soil is quickly depleted of nutrients. Biodiversity is then rapidly reduced in that area.

桉树被种下后，就进入了快速收获和重新种植的循环。结果，土壤养分会被迅速耗尽，生物多样性也会急剧下降。

Bamboo is an excellent source of fibre for making paper, especially for water-absorbent paper like toilet paper and tissue paper.

竹子能提供优质的造纸纤维，特别适合用来生产像卫生纸和纸巾这样的吸水性纸。

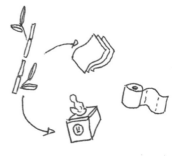

Tapirs resemble pigs and have trunk-like snouts. The tapir is an herbivore and can weigh between 150 and 300 kg. Tapirs are closely related to horses and rhinos.

貘外表像猪，鼻子像大象鼻子，它是一种植食性动物，体重在 150 千克到 300 千克之间。貘是马和犀牛的近亲。

The tapir uses his extended, prehensile snout as a snorkel when swimming. The tapir is an extraordinary seed disperser; therefore local extinction of tapirs leads to a loss of biodiversity.

貘的鼻子能伸缩、缠卷，在游泳时可伸出水面换气。貘是勤劳的种子散播者，因此貘灭绝后会引起当地生物多样性的丧失。

If a structure is flexible enough, it can move along with the force of an earthquake and absorb some of its impact. Can you do that?

如果一种建筑物有足够的韧性，就能在地震时随着力的作用晃动，并化解掉一些冲击力。这种事你能做到吗？

Would you like to have a tapir as a pet, or would you prefer a pig?

如果可以养一只宠物，你喜欢养一只貘还是一头猪？

通过种植桉树来获得造纸纤维的做法是否合理？

Does it make sense to plant eucalyptus trees as a source of fibre to make paper?

Bamboo not only helps hold water and regenerate soil, it also provides building materials in the form of poles. It can also be used to make paper, while increasing biodiversity. Would you say that bamboo is a generous grass?

竹子不仅能储存水分、促进表土再生，还可以提供建筑材料、造纸原料，以及提高生物多样性。你认为它是一种乐于奉献的草吗？

Do It Yourself!

自己动手!

Let's find out how easy it is to make paper or cardboard. Take a newspaper and remove all the inserts printed on glossy paper. Soak the newsprint in a little water for a day or two. Stir it until the paper had completely dissolved in the water. Now spread a thin layer of this pulp onto a cloth and let it dry. You will now have a thick sheet of paper or a thin sheet of cardboard.

让我们来看看造纸或硬纸板有多么容易。取一张报纸，剪掉上面的蜡光纸，用少量水浸泡一两天，搅拌直到完全化开，然后把纸浆倒在一块布上铺成薄层，晾干后，你就能得到一张厚纸或薄纸板了。

学科知识

Academic Knowledge

生物学	貘是食草动物，它的摄食和排泄对植物生长有重要的促进作用；貘有很好的潜水能力，也吃水生植物；当受到美洲豹或美洲狮攻击时，貘会逃进密林中的狭窄通道里躲避，低垂的树枝会不断刮擦攻击者的脸，使它们很难捉到貘；桉树大约有700个品种，都是澳大利亚本地物种；人们发现了约200种海里康属的外来热带植物，它们耐强降雨但不耐干旱；海里康属植物是蜂鸟的重要食物来源，蜂鸟和一些种类的蝙蝠几乎是它们专一的授粉者。
化 学	桉树分泌一种能杀菌的萜类化合物，弥漫在树的周围，以此来对抗竞争；纤维素是一种聚合物。
物 理	竹子茂密的枝叶能使周围环境温度降低大约10摄氏度，从而营造出清新凉爽的空间，这是熊猫、老虎和貘喜欢的栖息地；桉树能吸收地表下10米处的黄金，并通过枝叶排放出来。
工程学	貘依靠伸展的鼻子和长长的上唇抓住树枝、剥离树叶，并且不弄伤树就可摘到水果；人们把纤维素转变成纤维，再把纤维转变成人造丝。
经济学	桉树是一种速生树种，用于生产木材、清洁油、天然杀虫剂、染料和药物（特别是治疗关节痛疼的药物）；蜜蜂采集桉树花粉酿蜜；桉树耐火；桉树不耐霜冻，只能种在无霜区。
伦理学	外来物种虽然能提高经济收益，但也会和本地生物争夺水，并危及生物多样性，这一问题值得思考。
历 史	在南美洲发现了始新世时期（5 100万年前）的桉树化石，在新西兰发现了中新世时期（2 500万年前至500万年前）的桉树化石；桉树成为澳大利亚当地的优势树种与大约5万年以前，原住民祖先的到来有关；1770年，詹姆斯·库克船长首次把桉树从澳大利亚带到了欧洲。
地 理	貘栖息在南美洲、中美洲和东南亚（马来西亚和印度尼西亚）的丛林里。
数 学	竹子中含有60%的纤维素；经过3年的生长，此后70年，每公顷竹林（根据每根竹子重60千克，每公顷竹林包含500根竹子来计算）每年可获得至少30吨的生物量，这一数字远远超过了桉树。
生活方式	在汉语、韩语和日语中，"貘"这个词是相同的，它在中国神话中被认为是一种能吃梦的神兽；竹子被哥伦比亚人和印度尼西亚人称作"植物钢材"。
社会学	eucalyptus（桉树）是一个合成词，eu的意思是"好"，calyptos在希腊语中的意思是"覆盖"。
心理学	推销员利用心理学知识来说服人们购买某个产品，包括赢得信任和使顾客的决定合理化，面对不断的拒绝，推销员不能气馁；《推销员之死》这部剧，说的是一个好心的推销员因为看不到工作的意义以及无力实现美国梦而崩溃的故事。
系统论	桉树和其授粉者——蜂鸟、蝙蝠和负鼠是共生关系，这些动物不受桉树分泌的酚类化合物的影响。

情感智慧
Emotional Intelligence

貘

　　貘很怀旧，喜欢回忆森林过去的样子。他观察敏锐，作出的批评言之有据，难怪他觉得受到了欺骗。当谈论种植竹子时，他很克制，但是当谈到海里康花时，他变得情绪化。他热衷于通过发掘竹子的更多用途，来提高它的经济价值和接受度。当他描绘竹子超越其他树种的大量商业用途时，他很享受松鼠的补充说明。他感兴趣的不仅是种竹子能造福人类这一点，而是由此展开的更宏大的场景：竹子对这座山以及对构成其栖息地的生态系统至关重要。貘具备自我认知能力，知道自己的需求来自他与这块栖息地中其他生物和有机体的生态关系。

松鼠

　　松鼠从祖母那里知道了这块栖息地过去的样子。他很难相信会有人认为引进外来树种是个好主意，除非这个人被巧舌如簧的推销员洗脑了。松鼠很有学问但很谦虚，当貘问他问题时，他很有耐心，也对貘表现出了足够的尊重。他思维活跃，对于竹子能造福人类这个话题，他热心于寻找答案，并为貘提供了思路。但他做得很隐蔽，并不是想炫耀自己的智慧。通过回答貘的问题，他也获得了一些新的认识。松鼠知道貘不仅是在思考这座山和种豆农民的处境，也是在寻求更好的栖息地。他支持貘去达成这个愿望。

艺术
The Arts

　　海里康花是一种美丽的热带花，几乎只由蜂鸟授粉。这种花有各种各样奇特的形状。找一些海里康花的图片，挑选三种你认为最稀奇古怪的形状画下来，然后考虑一下这种形状是否有利于传粉。

思维拓展
Systems: Making the Connections

貘在其栖息地的生态系统中扮演了最重要的角色，这意味着如果它的生存受到威胁，整个生态系统就会面临崩溃的危险。作为植食性动物，貘以各种水果和浆果为食，广泛散播其种子，因此它是热带森林生物多样性的维护者。它是一种温顺的动物，在进化中发展出了通过逃进密林来摆脱食肉动物这样的防御策略，以避免冲突。通过貘，我们可以了解到生态系统对关键物种的依赖性，以及生态系统内部是如何解决冲突的。

在一个生态系统中，人们要在种植本土植物和为了经济利益种植外来植物之间作选择。推销员会尽力宣扬他所销售产品的优势，并诱导人们忽略他人的理性观点。当错误决定所带来的不利后果开始彰显时，往往已经太迟了。正如这则童话所示，我们要先集合一切现有的信息和智慧，要依靠我们今天的知识以及前人的知识，来发现基于本地现有材料以及所需材料的一系列机遇。这能增加就业机会，确保物种有更好的生存机会，提高生物多样性，增加水的存储，提高美丽的共生植物的欣赏价值。我们作决定时不能只聚焦于某个产品，而是需要关注到整个生态系统，深入理解某个建议的优势和不足，以保有更大的选择空间。

动手能力
Capacity to Implement

推销员必须基于逻辑、事实和数据来说服别人。让我们来扮演推销员吧。分成两组，一方支持销售桉树种子，另一方支持销售竹子种子，然后双方清晰、简洁地发表意见来论证自己的观点，尽量不要涉及太多技术细节。确保每个推销员都有机会陈述他的观点。然后讨论有没有可能一个推销员总是对的，而另一个总是错的……我们是否可以得出一个很有意义的结论：在某个地方正确的观点在另一个地方可能是不正确的？

故事灵感来自
This Fable Is Inspired by

路易斯·米格尔·阿尔瓦雷斯·梅希亚
Luis Miguel Alvarez Mejía

路易斯·米格尔·阿尔瓦雷斯·梅希亚毕业于哥伦比亚国立罗萨里奥大学，获得遗传学硕士学位，并在卡尔达斯大学获得农业工程学士学位。他是研究竹林及其生态系统物种再生方面的专家，是卡尔达斯大学植物园的负责人，对生物多样性的动态、水和本地物种推动经济增长方面有深入理解。他相信即使森林被破坏了，依靠热带生物的力量也可以恢复原貌。

图书在版编目（CIP）数据

冈特生态童书.第三辑修订版:全36册:汉英对照 /
(比)冈特·鲍利著;(哥伦)凯瑟琳娜·巴赫绘;
何家振等译.—上海:上海远东出版社,2022
书名原文:Gunter's Fables
ISBN 978-7-5476-1850-9

Ⅰ.①冈… Ⅱ.①冈… ②凯… ③何… Ⅲ.①生态环
境–环境保护–儿童读物—汉、英 Ⅳ.①X171.1-49

中国版本图书馆CIP数据核字(2022)第163904号
著作权合同登记号图字09-2022-0637号

策　　划　张　蓉
责任编辑　祁东城
封面设计　魏　来李　廉

冈特生态童书
竹子的馈赠
[比]冈特·鲍利　著
[哥伦]凯瑟琳娜·巴赫　绘

隋淑光　译

记得要和身边的小朋友分享环保知识哦！
八喜冰淇淋祝你成为环保小使者！